HOW TO
MAKE
BAMBOO
FLY RODS

HOW TO
MAKE
BAMBOO
FLY RODS

GEORGE W. BARNES

WINCHESTER PRESS

Barnes, George W   1927-
    How to make bamboo fly rods.
    Includes index.
    1. Fishing rods—Design and construction. 2. Fly
fishing. 3. Bamboo. I. Title: Bamboo fly rods.
SH452.B37   688.7'9   77-6738
ISBN 0-87691-237-4

Published by Winchester Press
205 East 42nd Street
New York, N.Y. 10017

WINCHESTER is a Trademark of Olin Corporation
used by Winchester Press, Inc. under authority
and control of the Trademark Proprietor.

Printed in the United States of America.

Dedicated to my wife, Jane,
who (mostly) puts up with
my "foolishness."

CONTENTS

INTRODUCTION

Lf you have ever experienced the thrill of catching a beautiful fish on a fly you tied yourself, and reflected that your skill in understanding and imitating the patterns of nature had made it possible for you to outwit a wily adversary, then you may have the beginning of an understanding of the super thrill of making the catch with a fly rod made with your own two hands. Assembling a rod kit from ready-made blanks can let you approximate this thrill, but the ultimate is reserved for those who build their rods from "scratch."

The art of constructing split bamboo rods has been held in mystery and apparently little has been published on the subject. When I built my first rod, a thorough search revealed only a few magazine articles with limited information. Much of the process had to be learned through experimentation and trial and error. But one fact stands out: If you have the basic skills of woodworking, and know how to use simple tools, then you can build a rod far superior to any that can be obtained from today's commercial sources.

Advancements in modern materials such as fiberglass and carbon filaments have done much for the sport of fishing, especially in reducing costs; but even a fine commercial cane rod, now largely made by machine and impregnated with rosins, cannot match the action and feel of handcrafted split cane. If you are lucky enough to possess a bamboo rod from the era of commercial handcraftsmen, then you know this superb feel and action. There are few examples of this art remaining today, however, and increasing labor costs, mechanization and declining pride in craftsmanship have taken their usual toll. Now the only alternative is to build your own.

The final incentive to build my first rod, after several years of contemplation, came from a manufacturer's comment that it was impossible to build a cane rod at home. More than a few home-built rods with superb fishing action have proven this statement to be completely false. Hopefully, this book, with its step-by-step instructions, sources of material, and helpful hints in general, will provide you with the incentive and desire to produce a fine fishing instrument of your own.

The joy of building is almost as exciting as the fishing itself.

George W. Barnes
Harpswell, Maine
February 1977

HOW TO
MAKE
BAMBOO
FLY RODS

CHAPTER I

THE PLANING BLOCK— KEY TO ROD BUILDING

T he most important aid or "tool" used in hand crafting rods is a good planing block. It is this item alone that makes it possible to correctly hold the split cane strips while planing, scraping, filing and shaping to the proper central angle and dimensions.

Details will be limited to planing blocks for six-strip rods, since there are several advantages to this type of rod construction:

1. Only one set of 60° grooves is needed to produce finished rod strips.

2. Individual strip dimensions are one-half the finished dimensions of the completed rod blank.

3. Ferrules may be fitted without heavy cutting of cane fibers, since the circular section of the ferrule closely approximates the resulting hexagon cross section of the rod blank.

The one disadvantage of the six-strip rod, if it can in fact be considered a disadvantage, is that the completed cross section is bisected by three clevage planes created by the glue joints. With modern glues this is no longer a problem.

Before the advent of space-age adhesives, the five-strip rod was developed to overcome this disadvantage of straight glue joints from one side of the rod blank to the other. Five-strip construction, however, presents more than a few complications that are better avoided. Since the central angle of these strips is 72° rather than the 60° of six-strip construction, the two remaining equal angles of the strip are 54°. A pair of 54° grooves tipped 18° from the vertical are required in shaping finished strips, obviously increasing labor costs in producing a planing block. The dimension of individual strips can no longer be obtained by simply dividing finished rod dimensions by 2; a factor of 2.236 must be used. More outside fibers must be removed in fitting ferrules, thus reducing rod strength and increasing the difficulty of producing professional appearing ferrule windings.

If you plan to build more than one or two rods, the cost of having a steel or aluminum planing block machined is well worth the investment.

Most well-equipped machine shops will have a milling machine with sufficient bed length to construct a 48″ block as detailed in the accompanying drawing. With adequate care a block can also be produced with a shorter bed length by cutting grooves in two operations.

As shown in the detailed drawing, each groove has the same slope and only one setup with varying depths of cuts is required to machine a block.

Without question steel is the best material to use, since it is heavy enough to provide excellent stability while working rod strips. If aluminum or some other light metal is used, the block should be drilled and countersunk to allow fastening to the work bench. It should be possible to obtain an excellent steel block from a machine shop for $70 or $80. Some sav-

Planing blocks. From bottom: steel, aluminum, wood.

3

STEEL PLANING BLOCK
FOR SIX STRIP ROD

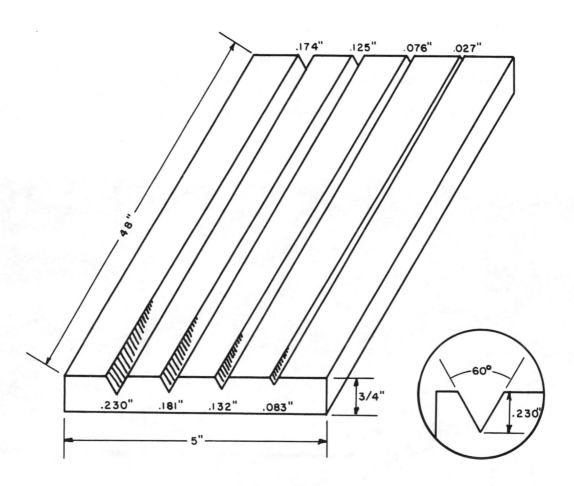

NOTE:

ALL 60° GROOVES CUT ON SAME
STRAIGHT SLOPE OF .056" IN 48"

CONSTRUCTION OF WOOD PLANING BLOCK
WITH SAW TABLE

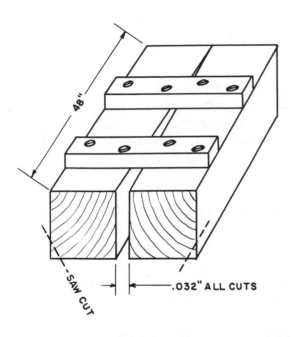

48"

SAW CUT

.032" ALL CUTS

SET UP FOR SAW CUTS

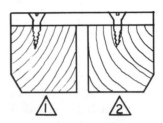

POSITION DURING
SAW CUT

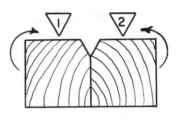

POSITION ASSEMBLED

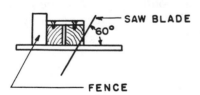

ASSEMBLED WITH
CARRIAGE BOLTS

SAW BLADE

60°

FENCE

AFTER SETTING UP FOR CUT, MAKE
CUT ON BOTH SIDES BEFORE CHANGING
SAW ADJUSTMENTS. REMOVE FASTENING
AND SET UP FOR NEXT PAIR OF CUTS
USING ONE STRIP FROM PREVIOUS CUT.

DIMENSIONS IN INCHES	
DEPTH OF CUT AT DEEP END	HORIZONTAL WIDTH OF SAW CUT AT DEEP END
.230	.133
.181	.095
.132	.076
.083	.048

5

ings may be possible by contacting a good-sized general contracting firm or well-equipped vocational school.

As with any tool made of steel, a certain amount of care will be required to keep the planing block in good condition. When the block is to be idle for any length of time, a good brushing with a stiff wire brush to remove all dust and cane particles from the grooves followed by a coating of light oil applied with a soft rag will prevent rust and deterioration.

Aluminum, of course, will not rust, but it nevertheless will oxidize. Since a stiff wire brush will scratch and mar the soft aluminum, possibly to the extent that very shallow grooves will be destroyed, a different cleaning process should be used. A stiff nylon brush or one made of brass should be suitable. Again, a thin coating of oil will tend to retard oxidation and preserve the life of the planing block.

If you plan to make only one or two rods (but don't count on this once you get started), a more than satisfactory planing block can be made from wood on a good saw table. A 4' length of clear kiln-dried hard-wood, such as maple or birch, should be obtained. A 2" by 12" cross-section will be adequate. With the blade of the saw table set in a vertical position, carefully cut the stock into five equal strips.

As detailed in the drawing on page 5, temporarily assemble two strips with scraps of ½" plywood so that they touch at one end and have a gap of 0.032" at the other; use a feeler gauge to control the dimension of the gap. Set the blade of the saw table for a 60° cut and very carefully set the saw fence to take a horizontal cut of 0.133" at the end of the assembly with the 0.032 gap. Again a feeler gauge can be used to mark the wood for the cut. Carefully take a cut of this dimension from both sides of the assembly.

Disassemble this pair of strips, rotate each piece 180° around its long axis and place them in contact along their entire length. A straight tapered groove with 60° central angle will result.

Using one of the pair of strips already cut and one uncut strip, make another temporary assembly with plywood gussets and the same 0.032" gap. This time reduce the horizontal cut to 0.095". Make sure that the temporary assembly is such that a second cut will result on the same surface of the previously cut strip of hardwood stock. Again make a cut on each side of the assembly. Disassemble, rotate and place in contact and a second but shallower 60° groove will result. Repeat the process two more times maintaining the standard 0.032″ gap but using horizontal cuts of 0.076″ and 0.048″ (see table of dimensions in drawing).

Once all five strips have been sawed and placed in contact, they should be securely clamped in their final position before drilling and fastening with a minimum of three ½″ diameter carriage bolts.

In making this type of planing block, I have found that there is a tendency to overcut on the horizontal dimension, which of course results in increased depth of the 60° grooves. Should this happen and result in grooves too deep for the rod under construction, it is a simple process to reduce the depth of the grooves by running the completed planing block through a jointer or thickness planer to remove some of the surplus wood from the surface of the block, thus producing shallower grooves.

Whether you decide to use wood or steel, once the planing block is completed it should be marked at 6″ intervals, since this is the basic spacing used to check dimensions of individual rod strips as they are planed, scraped, filed and worked to finish dimensions. On wood this can be easily accomplished with a soft pencil and try square. On steel the cross markings will have to be scribed with an awl or another similarly sharp pointed tool.

Wood planing blocks can be made on a shaper or with a hand plane modified to cut a 60° groove; however, a saw table is usually available in most home shops and this method is recommended if a machine shop is not to be used.

If the plane method is to be used, it will probably be necessary to have the plane iron modified at a machine shop unless machine tools are available, since the 60° angle is critical to rod construction. Slope of the grooves can be controlled by eye, since the taper of each fly rod is controlled by dimension of individual strips and not the planing block. A straightedge should be used, however, to control the plane and insure that grooves are not destroyed by the plane "running wild."

CHAPTER II

HAND TOOLS
AND
EQUIPMENT

Bamboo, or cane, as it is more correctly called, does not cut easily and dull tools have a tendency to lift and tear it. Cutting tools should literally be sharp enough to shave with. For this reason alone good quality sharpening equipment is essential. An ordinary coarse/fine oilstone, which can be found in almost any hardware store, should be obtained for preliminary sharpening and shaping of cutting edges. I have found a final sharpening operation on an India combination bench stone, first on the coarse India and then on the fine, will produce a more than satisfactory cutting edge on both plane and scraper. Woodcraft Supply Corporation, 313 Montvale Avenue, Woburn, Massachusetts 01801, has an excellent stone of this quality made by Norton Abrasive.

An ordinary 5″ or 6″ block plane has proven the most practical plane for me. It is easily held and controlled during planing of individual rod strips and is not so heavy as to produce fatigue. Small model-making planes tend to cramp fingers and are difficult to hold in the correct relative position to the planing block.

Homemade scrapers such as were used in the yachting trade of a decade or two ago are indispensable in final shaping. They are not to be found commercially and a subsequent chapter is devoted to their construction.

Ten-inch mill bastard files with comfortable handles are used in filing exterior cane fibers to a normal cross section at the location of interior dams, shaping of individual rod strips, especially at the locations of dams or nodes in the raw cane, final shaping of fine tip sections and as a lathe cutting tool when turning ferrule and reel fitting seats.

A 10″ rat-tail file will also be useful in shaping the interior of straight drilled cork grips.

A 60° center gauge will be helpful in checking the central angle of individual rod strips and the dimension of each side of the equilateral triangle formed during shaping. As time goes on and you become more experienced you will find that you will use the center gauge less and less as your eye becomes trained.

An old hunting knife with a relatively stiff blade is a must for halving raw cane sections and splitting individual rod strips.

A wooden mallet, either homemade, if you have a wood turning lathe, or commercially produced, is also indispensable in halving cane and in starting the knife when splitting individual rod sections.

A good quality micrometer reading down to .001″ is probably the most expensive hand tool involved in rod construction, but one that must be used almost continuously during final shaping of individual rod strips to guarantee that the finished rod will conform to planned cross-sectional dimensions, feel and action.

Several simple rectangular wood sanding blocks sized to accept a ⅓ sheet of sandpaper will also be found helpful, both in preparing individual rod strips and final finishing of completed rod blanks.

A selection of good camel hair brushes (water color brushes) will be needed for varnishing rod sections and applying color preservative and varnish to guide windings.

A good hacksaw with fine blade is probably the best type of saw for cutting rod strips to length.

A regular carpenter's gouge or a gouge designed for a wood turning lathe should be on hand to remove dams or nodes from the interior of halved canes. The gouge handle in any case should be substantial, as several blows of a wood mallet will be required to break dams loose.

A carpenter's rule (I prefer a steel tape type) will be needed for measuring lengths of rod strips.

As described in the separate chapter on homemade scrapers, a burnishing tool will be needed to "turn" the scraper edge. This tool can be made by grinding down a three-cornered file or can be obtained from Woodcraft Supply Corporation.

To avoid painfully cut fingers, a pair of good serviceable leather gloves will be needed for splitting, planing and scraping operations. They will also be handy during straightening of the individual rod strips.

A common propane torch with both a pencil flame and a regular flame tip should be obtained. The pencil flame will be used for straightening individual rod strips and the regular flame tip for heat treating strips before assembly.

Hand Tool and Equipment Check List

        Regular quality coarse/fine oilstone
        India combination bench stone (coarse/fine)
        Block plane
        Homemade scrapers (see separate chapter on their construction)
        Two 10″ mill bastard files
        One 10″ rat-tail file
        60° center gauge
        Stiff-bladed hunting knife
        Wooden mallet

Essential tools. Clockwise from bottom: 60° center gauge, knife, block plane, mallet, mill bastard file, micrometer and halving tool. Scraper is in the center.

Micrometer
Simple wood sanding blocks
Assorted camel hair brushes
Hacksaw with fine blade
Gouge
Carpenter's rule
Burnishing tool
Leather gloves
Propane torch with pencil flame and regular flame tip

CHAPTER III

MAKING,
SHARPENING,
AND TURNING
A SCRAPER

T he scraper I have had great success with cannot be obtained commercially. It is handmade and was widely used in the yachting industry during the heyday of that business. I was introduced to it by my father, who spent his early years as a yacht captain. If you can find an "oldtimer" who went yachting, he will be able to show you how to make a scraper in nothing flat, for all yachtsmen knew the secret. Otherwise, you will have to be guided by the following description and trial and error.

Good steel for the scraper can be cut from an old hand saw, but it is a long and tedious task to cut the steel with a cold chisel. Once cut to rough rectangular shape, the edges will have to be ground to final shape on an emery wheel, as the cold chisel will produce an irregular cut. Very suitable steel can be obtained from Woodcraft Supply Corporation. These steels are rectangular cabinet scrapers approximately 2½″ × 5″

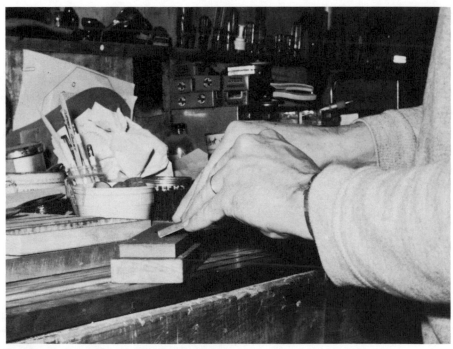

Sharpening scraper on fine bench stone.

16

made of Sheffield steel and are ordinarily sharpened square and then burnished on each edge of the square face. This steel is too flexible for heavy scraping but is more than adequate for scraping the small surfaces that are encountered in rod construction.

The first step after obtaining or cutting the steel is to grind a bevel on the edge to be used for cutting. The bevel should be approximately 30° and is best produced by very careful grinding on an emery wheel, but can be done with a good new mill bastard file.

Once the bevel is shaped, it must be sharpened and, like all tools used in cutting Tonkin cane, must be honed to a very keen edge. I prefer to start with a fairly rough bench oilstone and work down through several grades to a very fine stone. You should literally be able to shave hair off your arm before attempting to "turn" the scraper. Care must also be taken

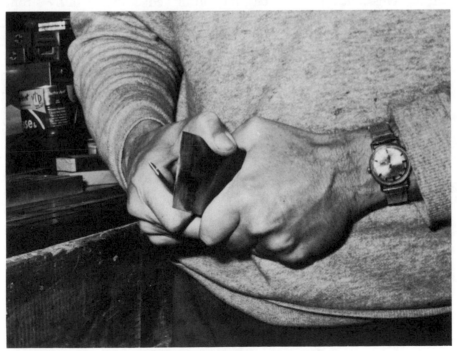

Turning scraper with burnishing tool.

that the bevel is all on one face of the scraper and that any sharpening done on the back of the scraper is done with it flat on the stone. As in any sharpening job, a wire edge can be avoided by moving the cutting edge toward the stone when sharpening.

Much better results will be obtained if a "cutting" oil is used with the stone. Light lubricating oils such as "3-in-1" just do not float cuttings away sufficiently to keep the stone from clogging. Rather than go to the bother of stocking special sharpening oil, I prefer to make my own. A mixture of ⅓ No. 10 motor oil and ⅔ kerosene or No. 2 fuel oil by volume will give excellent results. I had problems with sharpening for many years before I finally stumbled onto the fact that it was my oil and not my sharpening technique that was at fault.

Once a keen edge has been obtained, the scraper is ready to be "turned." With the scraper held firmly in the left hand, place a burnishing tool against the front side of the bevel at approximately 40–45° to the cutting edge and draw it along the cutting edge with a fairly strong pressure (see photo showing this process). This will roll or turn the cutting edge approximately 90° to the back of the scraper. Several passes of the burnishing tool may be necessary to sufficiently turn the cutting edge to the angle where it will scrape and cut easily. Commercial hand-burnishing tools are excellent for this purpose; however, if you do not wish to purchase this tool a very satisfactory one can be made by grinding down an old three-cornered file to the point where it is completely smooth, rounding the corners during the process.

When it comes time to resharpen the scraper (and this time will come often), the turned cutting edge is first filed off with a mill bastard file held flat against the back of the scraper and the sharpening and turning process is started just as if you were making a new scraper. It should not, however, be necessary to go back to the emery wheel for every resharpening.

CHAPTER IV

MAKING
SPECIAL TOOLS
AND AIDS

There are a number of tools and aids which should be built before actual rod construction is started. With the exception of the heat treating oven, all of these can be done without; however, experience has shown that they are all of great assistance in producing a good rod and will tend to make the construction even more enjoyable. Without doubt, all can be improved upon or made more elaborate, but, as pictured and described here, they are serviceable and will fulfill the intended function.

Heat Treating Oven

A good heat treating oven can be simply made from a 5′ length of 1½″ diameter black iron pipe. Each end is closed with a turned wooden plug which has a ⅛″ vent hole drilled through its entire length (see sketch on page 54). Smaller wooden plugs are used to seal the vent holes once heat treating is completed.

In use, the oven is suspended from the workshop ceiling in two wire loops so that it can be continuously turned during the heat treating process. If desired, shorter auxiliary wire loops can be used to store the oven closer to the ceiling when not in use, where it can then be used for storage of individual strips as the planing and scraping takes place.

Wrapping Vise

A wrapping vise to hold a freshly glued-up rod blank during wrapping is almost indispensable, unless you have someone willing to hold the "gooey" blank. The one pictured has proven a great aid and is constructed as follows:

A short length of 2 × 4 is mortised into a 12″ length of 2 × 10 stock to provide a stable base. If mortised in near one end of the base an auxiliary support can be fashioned at the other end of the base to prevent the rod blank from sagging while being wrapped.

The other, movable "arm" of the vise is also fashioned from a length of 2 × 4 and is attached with a 3″ butt hinge. Clamping action is provided by a simple cam, which is essentially a circular piece of plywood with a handle extension. A concentric semicircular slot is cut just inside the periphery of the circle and "rides" on a lag screw fastened to the fixed 2 × 4 upright. A second lag screw located "offcenter" and attached to the movable arm of the vise provides the necessary cam action.

Jaws are made by clamping two short pieces of pine together and boring

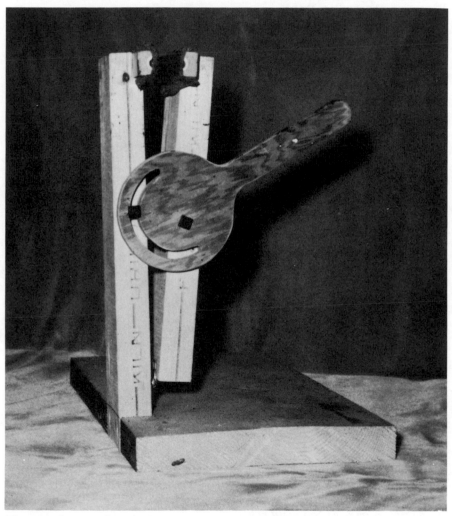

Homemade wrapping vise.

21

a ⁵⁄₁₆″ hole the length of the joint (a drill press is handy for this operation; however, it can be done with a hand brace and bit). Jaws are mortised into the vise arms to provide stability, and are covered with thin sheets of foam rubber or plastic to prevent marking the rod blank while it is clamped in position for winding. The sponge rubber used on the pictured vise was obtained from boxes used in shipping tree ripened oranges.

Pressure Wrapping Bobbins

Referring to the accompanying photograph showing both an assembled and "exploded" bobbin may be helpful in understanding the following description:

The frame of the bobbin is constructed of aluminum flat bar stock ap-

Homemade wrapping bobbins. Exploded view shows round-headed stove bolt used for bobbin axle to provide adjustable tension, aluminum frame and carpet thread tube from used ball point pen.

proximately $\frac{3}{16}'' \times \frac{3}{4}''$. Two right-angle bends are formed to make a "U" shaped frame with the opening slightly longer than the bobbin spool to be used. (Large spools should be used to provide sufficient capacity to hold enough heavy carpet thread to wrap at least one rod blank.)

The tube of the bobbin is made from a short section cut from a used ball point pen filler. A small tubing cutter should be used in cutting to the desired length to prevent rough edges which could cut the wrapping thread. A hole of sufficient diameter to accept the tube is drilled in the center of the "U" frame and is then slightly countersunk, using a regular wood countersink and drill press if available.

Using a center punch as a flairing tool, slightly flair one end of the tube before inserting into the "U" frame. The tube is held in place with a drop of solder applied as a collar and the tube is then solidly anchored with a few more taps on the flaired end. (Ordinary solder will not adhere to aluminum but will easily form a good collar on the brass used in most ball point pen fillers.)

The shaft on which the spool turns is an ordinary $\frac{1}{4}''$ round-headed stove bolt inserted through two holes drilled in the "U" frame for this purpose. Tension to the degree desired while wrapping rod blanks is provided by tightening the stove bolt nut. Washers shown in the exploded view are inserted between the spool and the "U" frame to prevent too much friction.

Those who do not wish to wrap their blanks by hand may wish to contact Cecil Pierce, Southport, Maine 04569, who has developed an excellent rod wrapping machine.

Finish Winding Jig

An excellent but simple winding jig, as pictured, can be made from a few scrap pieces of pine and plywood.

Pine uprights with slots for holding the rod being wound are screwed to a plywood base which in turn is fastened to pine cleats. The bottoms of the slots should be formed by drilling a $\frac{1}{2}''$ to $\frac{3}{4}''$ hole and then sawing the remainder of the slot to provide a circular cradle for the rod blank to turn in.

Three spool holders made from small dowels are provided for holding winding silk at several different and desirable angles to the tension devices. Tension devices (two are provided) are made from round-headed stove bolts, light springs and two hard-surfaced fiber faucet washers, available at any plumbing shop.

Bottom nuts on the tension device are inletted into the bottom of the plywood base so that a wrench is not needed during tightening. A second nut on top of the plywood base when tightened down provides good stability.

Winding silk is run from the supply spindle between the two fiber faucet

Homemade winding jig. Note light spring and two hard-surfaced faucet washers to provide adjustable tension when winding guides and ferrules. Two tension assemblies and spindles provide variety of winding positions.

washers and then to the rod section being wound. Tension on the winding silk can be varied as desired by simply turning the stove bolt to exert more tension on the light spring.

In use, two spools of winding silk are placed on the supply spindle, with the silk being used fed from the top spindle to bring it to the correct height to match the fiber washers applying pressure.

CHAPTER V

OBTAINING MATERIALS

Good quality Tonkin cane can be obtained from C. H. Demarest Company in New York. Even though import restrictions were lifted several years ago, cane is not always available on short notice and this is one of the first items that should be ordered. It is also advisable to check with Demarest on their current minimum order, as they cannot ship a single piece of cane. Minimum order used to be five pieces and this will provide enough cane for five or six good fly rods. If available, 1¾" to 2" diameter cane in either 6- or 8-foot lengths is preferred. Smaller diameter cane is acceptable, but will not provide as many individual rod strips per piece of cane as the larger diameter.

There are many types of ferrules available from a number of sources. E. Hille's nickel silver ferrules have proven more than satisfactory but are no longer available in all sizes. Hille has a number of other types of ferrules available, but my experience is limited to their anodized aluminum. This appears to be a good ferrule, but has not been tested through a full fishing season. The Orvis Company has an excellent and beautiful nickel silver ferrule; however, I have discontinued using this particular ferrule due to a tight fit between the male and female sections furnished that has required extensive hand working with crocus cloth for a proper fit. A search of catalogs from various supply companies will provide other sources of ferrules.

Reel seats can also be obtained in many styles and in various materials. E. Hille has a good economical anodized aluminum reel seat which is available with wood fillers to fit various sizes of rod butts. If you intend to make your rod a showpiece as well as a fishing instrument, Orvis has a beautiful walnut reel seat with excellent hardware and high-gloss finish.

Snake guides, tiptops, keepers and butt guides are available in various materials, either as independent items or in sets for fly rods. I prefer to obtain mine in sets in stainless steel; however, others may prefer blackened steel to eliminate reflection. These can be obtained from E. Hille, The Orvis Company or L. L. Bean.

Cork grips can be obtained in several ready-formed shapes or as cork grip rings, if you wish to shape your own grip. These are also available from both Orvis and Hille.

Winding checks to provide a finished appearance at the front of the cork grip are available in several patterns and materials. Hille has a good selection of anodized aluminum winding checks. Orvis has a very attractive check. Care must be exercised that the correct size is ordered. (Be sure to order a check that will fit the longest cross-sectional dimension of the rod blank at the exact location of the winding check, not the dimension across the flat portion of the cross section used in dimensioning rod blanks for construction.)

Pliobond liquid adhesive for gluing wood fillers in aluminum reel seats can be obtained at most Sears outlet stores and in various hardware stores. This same adhesive can also be used to attach ferrules to the rod blanks if care is exercised in turning the ferrule seats. I have used this adhesive on a number of rods, but prefer to use stick ferrule cement because of its quick set quality. E. Hille also lists Pliobond in their catalog. Stick ferrule cement can be obtained from The Orvis Company, L. L. Bean or E. Hille.

There are a variety of cements and glues available, ranging from natural glues to epoxies, that can be used for gluing up individual rod strips into the finished rod blanks. After some testing, I prefer U. S. Plywood resorcinol waterproof glue for a number of reasons. The two-part glue can be accurately measured by volume (an old ½ teaspoon measuring spoon will give sufficient volume of glue for most operations). It has a fairly long working life, sets in a few hours, has excellent bonding qualities and is really waterproof. In addition to this, it spreads easily with a stiff brush. Both glue and glue brushes can be obtained at most hardware stores or builder supply houses. The one drawback to resorcinol is its red color, which shows all imperfections in joints. Clear epoxy will of course eliminate this problem.

Good quality color preservatives and rod varnish can be obtained from Bean, Hille or Orvis.

Sandpaper in grades of 220 grit, 400 grit and 600 grit will be needed to get a good finish on the completed rod blank. Most hardware stores and builder supply houses will have 220 and 400 grit in stock; however, 600 grit may have to be obtained from these sources on special order.

L. L. Bean has a good assortment of rod bags, both two and three compartment in various lengths.

Rod winding silk and/or nylon is available from Bean, Orvis and Hille in various colors, both solid and variegated. Size A should be obtained for general winding purposes.

Addresses of the above-mentioned suppliers are as follows:

L. L. Bean, Inc.
Freeport, Maine 04032

Charles H. Demarest, Inc.
19 Rector Street
New York, New York 10006

E. Hille
The Anglers Supply House
815 Railway Street
P.O. Box 269
Williamsport, Pennsylvania 17701

The Orvis Company, Inc.
10 River Road
Manchester, Vermont 05254

All of the above have catalogs available, with the exception of C. H. Demarest.

CHAPTER VI

SELECTING A ROD

Complete dimensions for three different rods have been included on the following pages. Each chart of dimensions includes cross-sectional dimensions at various lengths along the rod sections, the locations of guides and decorative windings, ferrule sizes required, approximate weight of the completed rod and the number of the fly line that has been found to be most suitable for the given rod.

The copy of the 8′ commercial rod may be the most practical rod to start with, as the tip dimensions are not as fine as those of the other two. Dr. Merrill Haskell, a retired dentist from Northeast Harbor, Maine, who grew up with fly fishing in Maine, says this rod will lift a fly line off the water as easy as any rod he has ever used. My own experience with this rod has been delightful. It has good action, will produce a long cast and has caught a lot of fish.

The copy of the Thomas Browntone is a very light active rod and will produce unusual thrills with even medium-sized trout. The photographs showing the various steps of rod construction are of a copy of this Thomas rod that was made for Richard T. Carroll of Southwest Harbor, Maine. The Thomas Rod Company is no longer in existence and the only way to obtain a rod of this quality now is to construct it yourself, or be lucky enough to locate one of the originals that is for sale.

The copy of the Old Thomas Rod, as the name implies, is one of the earlier models produced by the Thomas Rod Company. Apparently, this rod was made before the advent of modern glues and the original rod is wound at approximately 1″ intervals over its entire length. This rod also produces a very light action; however, it is not as light as the copy of the Browntone.

Using the micrometer, it is also possible to obtain the dimensions of other rods and reproduce them in your workshop. In measuring rods for reproduction, there are several hints that will be helpful:

Basic measurements should be taken at the standard 6″ intervals. An additional measurement will be needed wherever a uniform taper is not maintained throughout the 6″ standard interval. The Thomas Browntone is a good example of where such additional measurements become

COPY OF 8' FINE COMMERCIAL ROD
WEIGHT = 4 5/8 OZ.
NUMBER 7 LINE
14/64 FERRULE

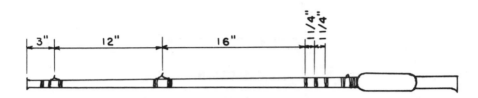

BUTT SECTION

	48"	42"	36"	30"	24"	18"	12"	6"	0"
FULL	.220"	.240"	.255"	.270"	.280"	.290"	.320"	.350"	.350"
HALF	.110"	.120"	.128"	.135"	.140"	.145"	.160"	.175"	.175"

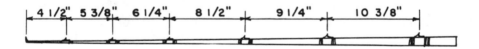

TIP SECTION

	48"	42"	36"	30"	24"	18"	12"	6"	0"
FULL	.062"	.085"	.115"	.140"	.165"	.185"	.200"	.210"	.220"
HALF	.031"	.042"	.057"	.070"	.082"	.092"	.100"	.105"	.110"

COPY OF THOMAS BROWNTONE
WEIGHT = 4 1/4 OZ.
NUMBER 4 LINE
11/64 & 13/64 FERRULES

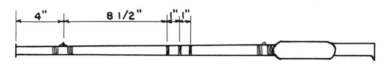

BUTT SECTION

	30"	24"	18"	12"	11"	10"	6"	0"
FULL	.208"	.223"	.246"	.315"	.341"	.388"		.388"
HALF	.104"	.111"	.123"	.157"	.170"	.194"		.194"

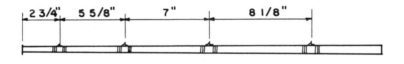

MID SECTION

	30"	24"	18"	12"	6"	0"
FULL	.165"	.169"	.172"	.196"	.203"	.206"
HALF	.082"	.084"	.086"	.098"	.101"	.103"

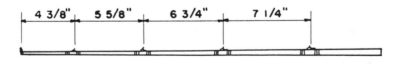

TIP SECTION

	30"	24"	18"	12"	6"	0"
FULL	.058"	.074"	.088"	.108"	.132"	.160"
HALF	.029"	.037"	.044"	.054"	.066"	.080"

34

COPY OF OLD THOMAS ROD
WEIGHT = 4 3/8 OZ.
NUMBER 5 LINE
11/64 & 14/64 FERRULES

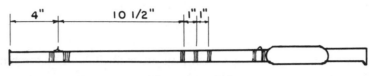

BUTT SECTION

	30"	24"	18"	12"	8 3/8"	6"	0"
FULL	.225"	.245"	.262"	.280"	.346"	.346"	.346"
HALF	.112"	.122"	.131"	.140"	.173"	.173"	.173"

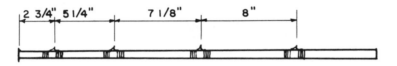

MID SECTION

	30"	24"	18"	12"	6"	0"
FULL	.170"	.179"	.196"	.204"	.204"	.225"
HALF	.085"	.089"	.098"	.102"	.102"	.112"

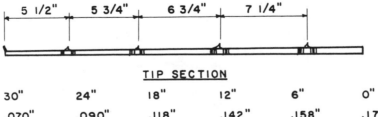

TIP SECTION

	30"	24"	18"	12"	6"	0"
FULL	.070"	.090"	.118"	.142"	.158"	.170"
HALF	.035"	.045"	.059"	.071"	.079"	.085"

necessary, for there is a short, quick taper immediately following the level section where reel seat and grip are attached to the butt section. (See dimensions of Browntone on page 34.)

All three dimensions of a hexagon cross section should be taken at each point measured and these should be averaged. Even in machine-cut sections, these dimensions may vary several thousandths of an inch.

Most rods have a level section with no taper at the end of the butt where reel seat and grip are installed. One series of measurements as close to the grip as possible should suffice. Obviously, some 6″ intervals will fall at snake guides or ferrules. When this occurs, a set of dimensions should be taken as close to the obstruction as possible.

CHAPTER VII

SPLITTING, ROUGH SHAPING, AND STRAIGHTENING

S plitting, rough shaping and straightening of individual strips is the first step in actual rod construction. The raw cane used to produce a rod blank must be somewhat longer than the rod section to be constructed. A 30″ rod blank will require a minimum of 36″ length in the raw cane, and a 48″ section will require a minimum of 54″. If cane has been obtained in 6′ lengths, these may be sawed in two if a 30″ rod blank is planned, but a full 6′ piece of raw cane will be required for a 48″ rod section.

Using an old knife and a wooden mallet, split the raw cane its entire length. Usually cane will have a natural split its entire length, or at least a good portion of its whole length, which occurred spontaneously as the raw cane cured. When splitting the cane into two sections, this natural split should be used as one of the cuts.

Once split, the half cane is laid flat on the work bench with one end against a stop block nailed solidly to the bench. The interior dams or nodes are then removed with mallet and gouge. Several sharp blows of a fairly heavy mallet will be required to break the dam loose. Once it is broken away, the interior of the cane can then be worked smooth, using the gouge as a hand tool; however, the gouge must be extremely sharp.

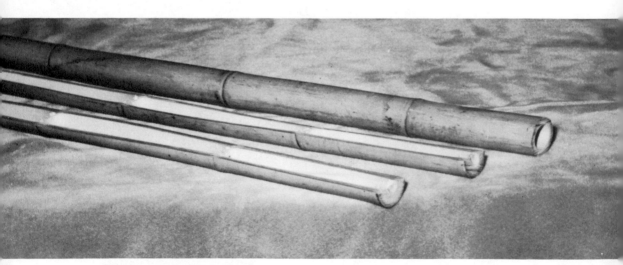

Whole cane and halved cane.

Halving cane.

As soon as the nodes are removed from the interior of the halved cane, the ridges on the exterior of the cane, which occur at each node location, are filed smooth with a 10″ mill bastard file. Usually a greenish fiber will be revealed as soon as filing is started. These are not the exterior fibers of the cane, which must be retained for maximum strength, so the ridges should be filed down to provide a normal cross section of the halved piece of cane.

The ends of the halved cane are then marked with a soft pencil as a guide for starting the splitting of the individual strips. A halved cane will provide the six strips necessary for the finished rod blank for most rod cross sections. However, the butt end of the raw cane should be used for the butt section of the rod to insure adequate thickness of the cane.

Removing interior dam with gouge and mallet.

The exterior fibers of the cane provide the strength and flexibility for the finished rod, since they are more dense, and it is desirable that the less strong interior fibers be worked away during planing and scraping.

Smoothing interior dam with gouge.

Individual rod strips are split off with the knife and mallet in much the same way as the cane was halved. Using the marks on the end of the cane as a guide, start the knife with a blow of the mallet and then work the individual strip free by prying with the knife and pulling the individual strip away by hand. It is a good idea to wear a substantial pair of leather gloves during this process, as some splintering may result.

As soon as all six strips have been split they are laid flat on the work bench or planing block with exterior fibers up. Although the dams or nodes are undoubtedly the strongest part of the growing cane, they will be the weakest part of a rod strip due to the irregular grain involved and the amount of fibers that must be cut away to obtain the correct cross section. Because of this, the individual strips must be so located that no two strips have a dam or node in the same location. With the six strips lying on the bench or planing block as described above, slide each suc-

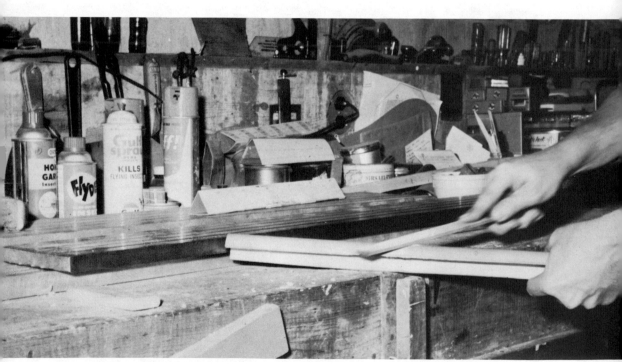

Filing outside of dam smooth.

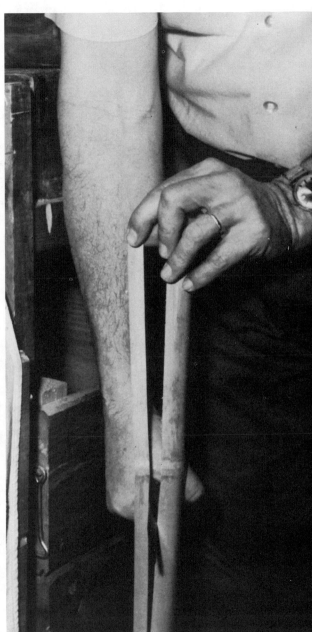

Starting to split section of rod.   *Splitting section of rod.*

cessive strip so that the dam or node comes in a different location. With most cane, an interval of 1″ to 1½″ on each successive strip will provide the correct staggering so that no two dams will occur at the same location in the completed rod blank. The maximum possible spacing available should be used.

As soon as the strips have been staggered, it is advisable to hold them in place with a heavy metal weight, since at this point in construction the strips will be warped and crooked. Using a try square and the butt of the one strip which was not staggered, all of the strips are marked at the butt end. Using a hacksaw with a fine blade, cut all of the butts along the line from this marking operation and then cut the individual strips to finish length, making sure that each strip is the correct length. Some rod builders advocate cutting the strips somewhat longer than the finished rod blank, but I have found that this is not an advantage during planing operations.

Split segments with dams staggered for cutting to final length.

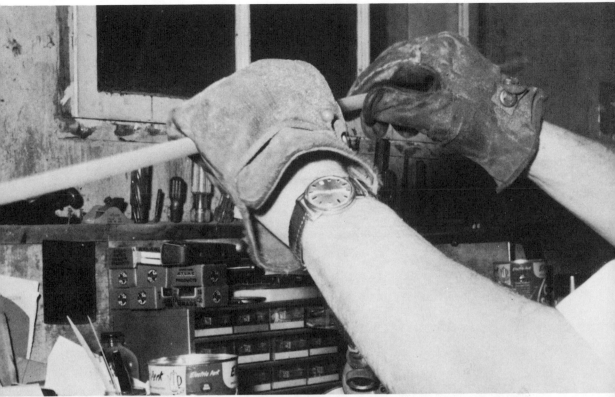

Heating with propane torch to straighten segments.

Straightened strips (with charred spots from heating), and others awaiting straightening.

Shaping to rough triangle. Rough shaping with knife aids in final finishing, as shape approximates grooves in planing block.

Using the planing block as a reference straightedge, lay a strip on the block with the exterior fibers down in order to locate areas where the strip must be straightened. A propane torch with a pencil tip flame is used to heat the cane for straightening.

Holding the cane on either side of the point where straightening must be accomplished, rotate the cane in the flame of the propane torch until it becomes fairly hot. At this point the cane will be soft and flexible and it is immediately removed from the flame and straightened by bending by hand and holding in the straightened position until the cane cools.

As soon as one spot has been straightened the cane is again laid on the planing block to locate the next point needing straightening and the process is repeated until a relatively straight cane results. During the heating process, interior fibers of the cane will be scorched and burned; however, this presents no problem, as they will be planed away during the shaping and scraping process.

Once all individual strips have been straightened, they should be cut to a rough triangular cross section with the knife to aid in holding them in the planing block at the start of planing operations.

Some rod builders advocate heat treating the individual strips at this point in rod construction. I have found that it is much more practical to heat treat after the strips have been worked to final dimension. Almost invariably, an individual strip will be destroyed during the planing operation, and if heat treating is done at this point in construction another strip will have to be cut, straightened and heat treated before a replacement strip can be shaped. With heat treating reserved for finished individual strips, no delay is experienced if a strip must be replaced. Heat treating, of course, tends to harden and strengthen the cane, which is the obvious reason for heat treating. The untreated cane tends to plane and shape easier than that which has been subjected to heat.

CHAPTER VIII

PLANING, SCRAPING AND SANDING STRIPS

P lace a straightened rod strip in the largest groove of the planing block, making sure that the butt end of the cane is at the deeper end of the groove and that exterior fibers are located against the side of the groove. Holding the blade of the block plane parallel to the surface of the planing block, take one or two cuts the entire length of the straightened rod strip. Then rotate the cane in the groove so that the exterior fibers lie against the opposite side of the groove and take an-

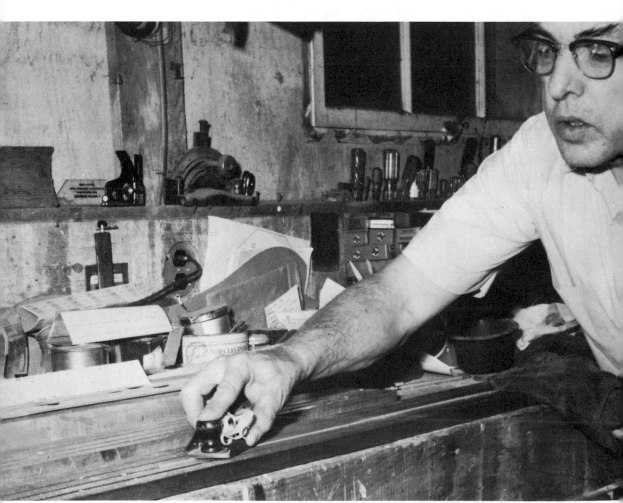

Planing a strip to rough shape (note glove to protect hand holding strip).

other one or two cuts the entire length of the cane. (See the accompanying sketch.)

As this process is repeated over and over again, the strip will magically assume a triangular shape with a 60° central angle. The central angle may be checked with the center gauge, which will also indicate whether or not the two legs of the triangle or sides of the strip are of the same dimension, as they must be. As time goes on and more experience is gained in planing the strips, this step can probably be eliminated as the 60° angle is virtually an automatic result of planing and any differences in the dimension of the legs of the triangle will be obvious to the eye.

The planing process is repeated, using successively smaller grooves in the planing block until the final cross-sectional dimensions of the strip are approximated. At the beginning of planing operations, it is advisable not to try to plane to exact dimensions. Better success will be obtained by stopping while the strip is $\frac{1}{64}''$ oversized and completing the shaping with a scraper and file.

A glance at the table of rod dimensions will immediately show that the completed rod blanks do not have a uniform taper from butt to tip but are in fact made up of a series of tapers and in some areas a uniform cross section. For this reason, I prefer to plane a rod strip to rough shape its entire length and then make a separate planing and scraping operation to obtain final dimensions. With the planing block marked at 6″ intervals and rod dimensions specified at the same interval, this becomes a relatively simple procedure.

Again beginning at the butt end of a strip and working towards the tip, work each successive 6″ increment to its final dimension with plane and scraper, with frequent checking with the micrometer. I have found that it is much more practical to preset the cross-sectional dimension on the micrometer and use it as a gap gauge rather than trying to turn the micrometer down onto the knife edge of the triangle cross section. Frequent checks with the micrometer are advisable, as the finish dimensions of the rod strip should not vary more than .001 inch from those specified in the rod chart.

A folded cardboard triangular prism with rod blank dimensions, similar to the one shown in the accompanying sketch, is a handy aid during planing and shaping of the individual rod strips. This can be placed near the planing block, is easily visible and eliminates the need to refer to books or notes.

Scraping to final dimensions.

As soon as final shaping of each individual strip is completed, a light sanding with 220 grit paper on a sandpaper block will remove any loose fibers. Completed strips should be stored in a cardboard tube or the heat treating oven to prevent damage until they are ready to be assembled.

As soon as all six strips are completed, they should be roughly assembled for heat treating. This can be easily accomplished with short lengths of masking tape, as explained later. The roughly assembled rod blank is tied with heavy carpet thread before heat treating and the masking tape is removed to prevent a sticky mess from melted adhesive.

During planing the plane blade must be kept extremely sharp and fine cuts must be taken as the section nears its final dimensions. Too heavy a

Checking dimension with micrometer used as gap gauge.

FOLDED CARDBOARD TRIANGULAR PRISM WITH
ROD BLANK DIMENSIONS IS A HANDY AID
DURING PLANING AND SHAPING

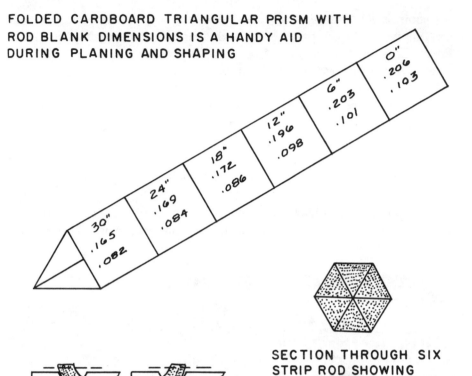

SECTION THROUGH SIX
STRIP ROD SHOWING
CLEAVAGE PLANES – NO
PROBLEM WITH MODERN
GLUES

CUT 1   CUT 2

SECTION SHOWING HOW CANE IS
ALTERNATED IN PLANING BLOCK
GROOVE DURING PLANING

SECTION THROUGH BLACK IRON
PIPE HEAT TREATING OVEN
SHOWING TURNED WOODEN
PLUGS AND VENT HOLES

EXCELLENT ROD CASE CAN
BE MADE FROM 1 1/2" PVC
PIPE AND FITTINGS WITH
SCREW PLUGS. DRILL TO VENT.

cut or the least dulling of the blade will result in the bamboo tearing and probably spoiling the rod strip. Heavier cuts may be taken at the beginning of each planing operation, but the best word here is caution.

Again, it is advisable to wear a heavy pair of leather gloves during planing operations, as the cane tends to become knife sharp and even a slight slip will result in painful cuts. An alternative to gloves, which at best are awkward, is a folded piece of sandpaper, which will grip the individual strip when minimum pressure is applied.

Where nodes or dams are encountered in the length of the individual strips, planing and scraping will ordinarily be more difficult. When this occurs, the dam should be worked down with the mill bastard file which will produce a good smooth surface and desired cross section.

The tip of each rod strip may also require final shaping with a file in lieu of the scraper, as the extremely fine dimension tends to tear when scraped. Be sure, however, to work the file toward the tip end of the section. I have spoiled more than one strip by not following this advice; the file caught on the rod tip, breaking it completely off.

Undoubtedly this sounds like a never-ending procedure to someone just starting his first rod. Let me assure you that it is nowhere near as complicated as it sounds, and even your first rod should not take more than a total of 50 or 60 hours from raw cane to completed fishing instrument.

CHAPTER IX

HEAT TREATING

As pointed out previously, I prefer to heat treat cane after all individual strips have been worked to final dimension rather than treat rough split cane and take the chance of having to go through heat treating for a single strip if one is inadvertently destroyed or does not meet required tolerances.

The individual strips should be assembled into their final position in the completed rod blank before treating. The easiest way to accomplish this is to lay a short piece of masking tape, sticky side up, on a flat surface

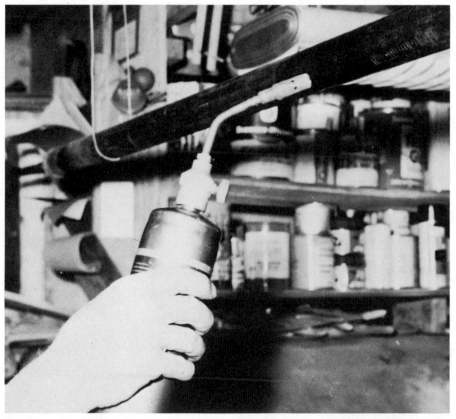

Heat treating strips. Pipe heating oven is suspended in wire loops. Flame from propane torch is passed back and forth along entire length of oven, with oven given a one-eighth turn every pass down and back. Individual strips can be heat treated in about one-half hour.

such as the back of a metal planing block. The outside fibers of each individual strip are then "stuck" to the masking tape with each successive strip just touching the adjacent one. Once all six strips are positioned, it is a relatively simple process to roll the strips into final position at the butt end. Other short pieces of tape can then be applied at intervals along the rod to temporarily complete assembly of the rod blank.

Once assembled, the blank should be tied with carpet thread at 3- or 4-inch intervals for its entire length. BE SURE AND REMOVE THE TAPE BEFORE HEAT TREATING OR A "STICKY MESS" WILL RESULT.

With the pipe heat treating oven suspended in wire loops at least a foot below the workshop ceiling, the preassembled rod blanks are placed in the oven, wooden plugs are inserted and the small vent plugs are removed. Heat is applied to the oven with a propane torch equipped with a regular flame tip by "walking" the flame back and forth the entire length of the oven. At completion of each "back and forth" trip, the oven is given a ⅛ turn to insure equal heat distribution, to reposition the rod blanks within the oven and to prevent the occurrence of hot spots and burned rod blanks.

After this process has been continued for about 15 minutes, steam should be emitting from the small vent holes in the wooden plugs and a distinct odor of heated bamboo should be evident. The process should be continued for another 15 minutes before heating is discontinued and the oven allowed to cool. During the initial stages of the cooling process, the oven should be given a ⅛ turn at frequent intervals, again to prevent scorching of the rod blanks.

This is probably the longest half hour in the entire rod construction process, since the next step will produce a glued-up rod blank where the fruits of your labor will really begin to show. Frequent checking of your watch will not really speed the process but will be unavoidable. Be assured that the process will eventually end.

I have recently been advised by another rod builder that this process can be speeded up somewhat by using smaller diameter copper wire. Copper, of course, is an excellent conductor of heat as well as electricity. Because of this, more care may be required in turning the oven to prevent scorching the rod. Copper pipe of ¾" to 1" inside diameter should be of adequate size to hold almost any fly rod cross section. Ordinary copper pipe caps can be used to close the pipe oven, but a vent hole will have to be drilled in each cap. Little will be gained from trying to solder one of the caps in place, as temperatures during heat treating will probably be high enough to melt the soldered joint.

CHAPTER X

ASSEMBLING
AND
GLUING STRIPS

A flat hard smooth surface is essential in assembling the individual strips in their final location in the rod blank. This can easily be provided by turning the metal planing block upside down and using the reverse side as a work surface.

No matter what work surface is used, it should be covered with ordinary household waxed paper to prevent excess glue from adhering to the work surface.

Six or eight lengths of masking tape approximately 1½″ long should be cut and placed in an accessible area before starting the assembly. Approximately ¼″ of each strip of tape should be doubled over to prevent adhesion and to allow easy removal in later steps of the work.

A single piece of tape is placed on the work surface, adhesive side up, and the first of the six strips to be glued is pressed onto the tape with the outside shiny cane fibers stuck to the tape and located so that the edge of the strip is at the very end of the tape. The remaining five strips are then positioned on the tape so that each strip is just touching the previous strip. This first piece of tape should be applied very close to the butt end of the rod blank being assembled and the butt ends should be carefully aligned.

With this single piece of tape firmly attached to each of the six individual strips, it is then a relatively simple matter to roll the six strips into their final location with the excess tape wrapped around itself to provide the necessary support. At this point, the individual strips may cross and almost entwine at the tip end of the rod blank. By working in 4″ to 6″ increments, it becomes an easy matter to correctly position the individual strips by gently rolling them between your fingers. As each increment is worked into its final position another strip of tape is added, making sure that the end of the tape is started on the same individual strip as the initial piece of tape. This process is repeated until the rod blank is taped its entire length.

Each piece of tape is then carefully unwrapped to the point where it is just touching the six strips and not wrapped onto itself. With the rod blank laying flat on the work surface, the individual strips can then be

Applying glue to finished strips in preparation for wrapping. (Note how triangular shaped strips are attached to masking tape with adjacent strips just touching.)

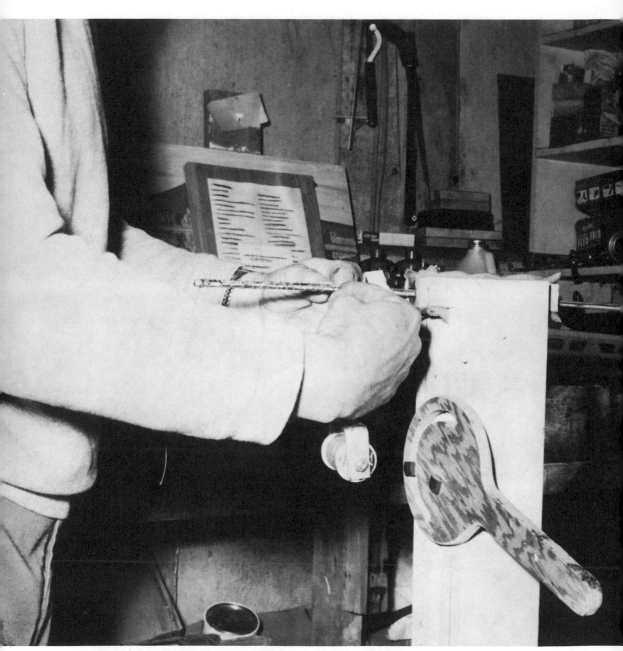

Wrapping rod blank while holding it in homemade wrapping clamp. (Rod blank is reassembled with masking tape at this step.)

separated with a thin knife so that each strip is laying flat on the work surface. At this point, the base of each triangular strip will be flat on the work surface with the apex exposed.

Using an old measuring spoon, mix a "batch" of resorcinol glue, carefully following the manufacturer's instructions. Using a stiff bristled glue brush, spread a thin coat of glue the entire length of all exposed individual strips, making sure that no area of the cane is missed. Resorcinol glue provides ample work time to complete assembly of the blank; however, time should not be wasted in completing the remaining steps.

Using the butt end tape, roll the six strips back together as they were in the temporary assembly in preparation for gluing. The remainder of the rod blank is then rolled into place just as before, and is held in position with the various strips of tape. The assembled rod blank is then inserted in the wrapping vise and carpet thread from both wrapping bobbins is tied at the butt end using clove hitches. When wrapping the mid-section of three-piece rods and tip sections of any rod, an auxiliary support attached to the back of the wrapping vise is a handy addition.

The rod blank is then wrapped its entire length with carpet thread which has been previously waxed to prevent glue adhesion, with the carpet thread crossed at ½" to ¾" intervals in much the same manner as lacing a boot.

The resorcinol glue will tend to make the cane feel soft and flexible and the rod blank will have a decided bow to it when the wrapping is completed. This bow and any "kinks" must be removed as soon as the wrapping is completed. This can be accomplished by rolling the rod blank back and forth on the flat work surface while applying gentle pressure, and by gentle bending at points of any extreme curvature. Final straightening can be accomplished by turning the planing block so that the 60° grooves are on the top surface and by gently pressing the wrapped blank into the grooves.

Too much emphasis cannot be placed upon the straightening process, since with modern glues it is virtually impossible to restraighten a blank once the glue has set. The straightening process is of course complicated by the fact that the carpet thread is wrapped its entire length, which makes it impossible to "sight" the blank for straightness.

In all of the rods I have constructed, there has been only one section which had an obvious bow when completed, so with proper care it is possible to obtain a good straight rod blank.

As soon as the blank is straightened, the ends of the carpet thread are tied together and a small wire hook is inserted in the end of the carpet threads. The rod blank should then be hung in a warm dry location where it can remain in a freely suspended vertical position for 24 hours.

As pointed out previously, clear epoxy cement may be used if the dark color of resorcinol is considered objectionable. I have seen several rods glued up with Elmer's 2-part epoxy where joints were practically invisible. I have not used epoxy in any of my rods and can offer no first-hand information on feel and action during actual fishing. Whether or not the same action results as with resorcinol will have to await the construction of a later rod.

As rods acquire some age they tend to darken, and the contrast with resorcinol's reddish color becomes less pronounced. I have not found the color of resorcinol at all objectionable. However, others may wish to eliminate the contrast.

CHAPTER XI

FINAL SHAPING, SCRAPING, SANDING AND TURNING FERRULE SEATS

A fter the glue used in assembling the rod blank has completely set, the carpet thread wrapping is removed. At this point, the outside of the rod will be covered with resorcinol glue, with ridges of glue apparent wherever the wrapping thread was in contact with the blank.

With the rod blank laying on a smooth hard flat work surface (again the bottom of a metal planing block is ideal), the glue and outside fibers of the cane are scraped off with the same scraper used in shaping the individual strips. Any irregularities at the location of the cane nodes are worked down with a mill bastard file so that each of the six surfaces of the rod blank are smooth and uniform.

The blank is then given a thorough sanding on each of the six faces with sandpaper held firmly on a wood sanding block. An excellent finish can be obtained by sanding in three steps using 220, 400, and 600 grit sandpaper (600 grit paper may have to be obtained on special order and should be acquired before this step in the rod making process).

As soon as the flat surfaces are completely sanded, the six angles of the rod section are slightly rounded with 600 grit sandpaper held in your hand without a sanding block.

The depth of each ferrule and the reel seat socket are carefully measured and these depths are marked on the completed rod blank. A simple but effective method is to use a fine dowel or metal rod to measure the exact depth of the ferrule and simply transfer this measurement to the rod blank.

The metal lathe I have access to is equipped with a four-jaw chuck. This type of chuck is of course not self-centering and some adjustment is usually required to center the rod blank before turning of ferrule seats actually begins. The blank is centered as closely as possible by eye. The lathe is then turned on and the tool rest is adjusted so that a pencil may be placed on the rest and advanced until the lead just touches the turning rod blank. A few turns will be sufficient to mark where the lead is in contact with the blank.

The lathe is then turned off and lead marks are examined. The jaws of the chuck are readjusted if marks are not uniform on the entire circumference of the blank, loosening the jaws where marks do not appear and

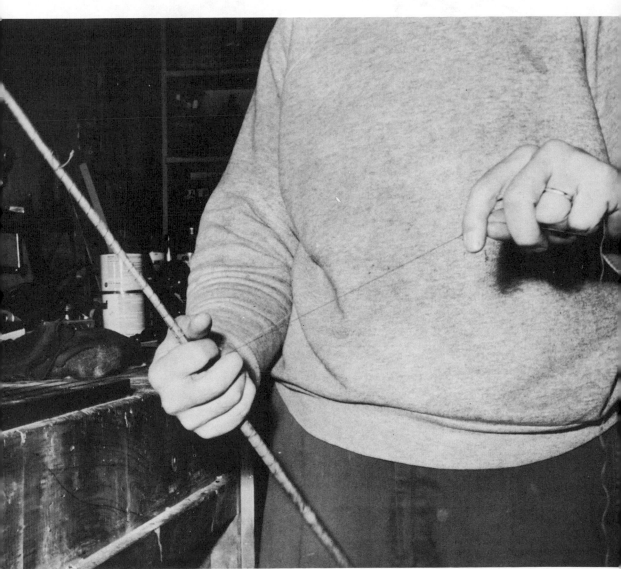

Removing wrapping thread from rod blank after glue has thoroughly set.

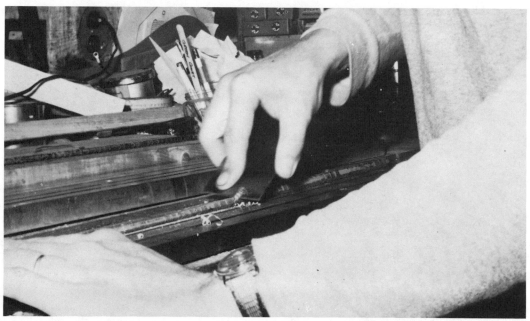

Scraping off glue with "turned" scraper. Outside shiny fibers are also removed during this step.

Filing nodes to normal cross section on glued up rod blank.

tightening those where marks are evident. The marking and jaw adjustment procedures are then repeated until marks are apparent on all angles of the hexagon section of the rod blank, extending completely around the circumference.

A three-jaw chuck will self-center the blank, but has the decided disadvantage of being more costly than a four-jaw chuck.

Sanding rod blank.

A word of caution: before turning the lathe on, the end of the rod blank extending through the hollow shaft of the lathe should be supported to prevent whipping. The simplest support is another person simply holding the rod end with his hand, grasping it loosely.

Although I have not done so as yet, it is my intention to make a simple lathe for this purpose. My plans include a suitable chuck attached to a hollow shaft of sufficient length to prevent whipping of the rod blank, two simple bearings for the shaft to turn in and a pulley on the end opposite from the chuck where power can be applied through a "V" belt. The entire assembly is envisioned as being similar to belt-driven mandrils available for grinding wheels. Pulley diameters will be selected to provide a relatively slow R.P.M.

Once the rod blank is assembled for turning and the length of the ferrule seat is marked as previously explained, the lathe is turned on, with rotation towards the operator, and the ferrule seat is turned to size by applying a mill bastard file to the blank. The simplest way to obtain the correct seat diameter is to go slowly and frequently try the ferrule on the seat. The correct diameter is obtained when the ferrule can be easily placed over the rod blank, but care should be taken that the fit is not so loose as to allow the ferrule to wobble on the seat.

Some wood turning lathes will have a hollow drive shaft of sufficient diameter to accept a completed rod blank. These lathes will usually have the drive shaft threaded to accept a face plate and it should be a relatively simple matter to obtain a four-jaw chuck to fit these threads, thus eliminating the need for a metal turning lathe. If a wood turning lathe is used for making the ferrule seats, the lowest speed available should be used.

CHAPTER XII

ASSEMBLING REEL SEAT, FERRULES, TIPTOP, GRIP AND WINDING CHECK

The next step is to glue the reel seat into position. If a reel seat with a wood core similar to that obtainable from The Orvis Company is used, it can be glued directly in place with resorcinol glue. If an anodized aluminum seat, which requires a wood filler, is used, it is advisable to glue the filler into the reel seat sometime in advance of assembly. Pliobond glue gives excellent adhesion between the anodized aluminum and the wood filler; however, the final assembly of wood filler to cane blank is still best done with resorcinol. In gluing the reel seat into position, care should be exercised that the point of attachment for the reel itself is carefully lined up with a flat surface of the rod blank.

If there is a slight bow to the butt section of the rod, the outside curvature of the bow should be the surface lined up with the point of attachment of the reel, since this will also be the surface receiving line guides.

Gluing reel seat in place (filler has been previously glued).

The cork grip is cut to the desired length (assuming that a grip will not be formed of individual cork rings) and, if the grip has not been taper bored, the hole in the grip is carefully worked with a rat-tail file so it can be pushed into place from the tip of the rod section with only slight pressure. Several trials will probably be needed to get the hole in the grip worked to its proper dimension.

Once it has been properly fitted, the grip is removed and its length is lightly marked on the rod blank to indicate its final position. A very thin coat of resorcinol glue is then applied to within ¼″ of the tip end of the grip location and the grip is reassembled for the last time. I usually use a preformed grip, available from most supply houses, rather than resort to shaping a cork grip myself. It is possible, of course, to turn your own grips if you wish to make as many rod components as possible.

Gluing cork grip in place.

If a wood turning lathe is equipped with a three- or four-jawed chuck and has a hollow drive shaft of sufficient diameter to accept the butt section of the rod, cork rings may be glued directly to the rod section for shaping the grip as an integral part of the rod.

I prefer to turn grips on a temporary mandril. The mandril can be turned from scrap material, leaving a fairly large diameter section to accept the drive spur of the lathe. The mandril itself is turned to a diameter slightly smaller than the hole in the cork rings to be shaped and somewhat longer than the overall length of the grip desired. A second turning is also made with a larger diameter. This is then bored to accept the mandril and acts both as a clamp for the cork rings and as a turning center for the cup spur of the lathe tail stock.

If care is used in the application of glue, cork rings may be placed directly on the smooth wood mandril. But I prefer to apply a layer of ordinary household waxed paper as insurance. Once the cork rings are placed on the mandril, the second increased-diameter piece is slipped over the end of the mandril and the entire assembly is mounted in the lathe. Pressure is applied by tightening the tail stock adjustment and the assembly is left until the glue has set. Cork may be glued with resorcinol, epoxy or Pliobond. If epoxy is used, the wrapping on the mandril should be of polyethylene film, since most epoxies will not adhere to this material.

Actual shaping of the grip should be done with judicious use of a fine wood rasp and fairly coarse sandpaper. Final finish is accomplished with finer grades of sandpaper, which tend to smooth rather than shape the cork. Once completed, the grip is then installed as described above.

This same process can be used to make a so-called skeleton reel seat where an extremely light rod is desired. In this approach, additional cork rings are glued up to provide the required length for the reel seat. The seat is turned to the correct diameter to accept reel retaining rings. The assembly is then removed from the lathe and the second oversized portion of the mandril is removed. Retaining rings are installed and an additional cork ring is glued in place to keep the retaining rings from falling off. The mandril is reassembled and placed back in the lathe with pres-

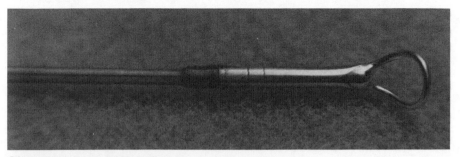

Closeup of tiptop and winding.

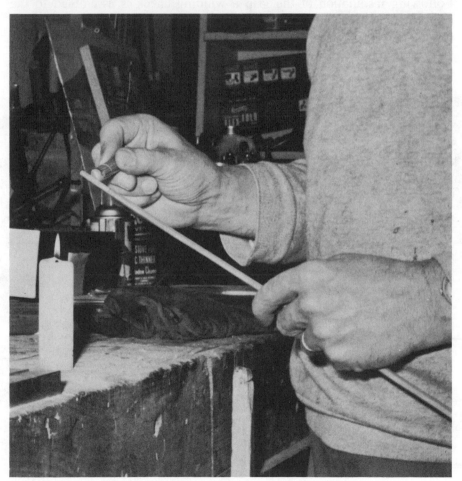

Applying ferrule cement to seat previously turned on rod blank.

sure applied until the glue has set. The last cork ring is then turned to the desired shape.

Very effective reel retaining rings can be made from short sections cut from an appropriate size of polyethylene pipe.

Where weight is not a prime consideration, I still prefer to use a walnut reel seat simply for its beauty.

Following installation of the grip a winding check is assembled to the grip to provide a finished appearance. In ordering winding checks it should be remembered that the diagonal dimension of the rod blank is

Cementing ferrule in place.

the dimension of the winding check and not the dimension across two flat surfaces used in constructing the blank itself. Anodized winding checks can be adjusted to larger size by judicious filing.

Ferrules and the tiptop are then assembled. Some people advocate using Pliobond glue for this assembly, but I have found that it does not work well with cane rods, even though it is excellent for fiberglass rods, and I prefer to use the old-fashioned stick ferrule cement.

When using stick ferrule cement, soften it in the flame of a candle or alcohol lamp and apply it to the previously formed seats in the rod blank. With the tiptop itself, it may also be advisable to work some ferrule cement into the socket of the tiptop before it is applied.

Holding the rod blank in one hand, and the ferrule in the other, gently heat the ferrule in the flame of the candle until the temperature is sufficient to remelt the ferrule cement on the seat. The ferrule is then quickly pushed into position on the seat and is held in position until the cement rehardens. A glove, of course, should be worn on the hand holding the ferrule to prevent what could be a rather nasty burn.

Surplus ferrule cement is then removed by scraping and the ferrule is wiped with a rag dipped in lacquer thinner to remove final traces of the ferrule cement.

At this point, you may have a strong desire to assemble the rod and see how it "feels." The correct reaction to this desire is "no," since the rod should not be assembled until the windings are applied to the ferrules to give the additional support needed.

CHAPTER XIII

VARNISHING

T he rod blanks should be varnished before adding snake guides in order to prevent runs and "lace curtains." Only a good grade of rod varnish should be used and it should be applied with a soft, good-quality camel hair brush. Varnish, of course, is flowed on, not worked back and forth as paint is.

The rod blank should be carefully wiped with a soft rag before the initial coat of varnish is applied. As soon as the blanks have received a coat of varnish they should be again hung in a vertical position in a warm dry location free of all dust. The varnish should be allowed to dry for 24

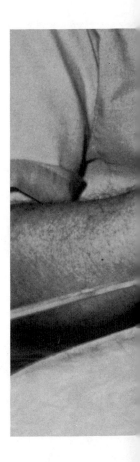

hours and the blanks should be thoroughly rubbed with very fine steel wool before another coat of varnish is applied.

If desired, your name, date of construction, rod specifications, etc., can be printed on the rod with India ink and a crowquill pen before the second coat of varnish is applied.

Three coats of varnish with a good rubbing between coats will produce a mirror finish. More than three coats of varnish should not be used, as any more will tend to detract from the rod action.

Varnishing rod section prior to winding guides. Name, date, etc., may be added in India ink after first coat is dry.

CHAPTER XIV

WINDING SNAKE GUIDES, KEEPER AND FERRULES

Windings of size A rod winding silk or nylon will be required to hold the snake guides and the keeper in position and to strengthen the rod at the ends of ferrules and the cork grip. The description which follows is detailed for winding the snake guides, but the same process is used on all windings. It will be helpful to refer to the sketches A through E, showing how windings are applied, and to the photographs showing the process.

Should the tip rod section have a slight bow when completed, snake guides should be installed on the outside of the bow.

With the snake guide attached to the rod blank in its final location with a strip of tape around one leg, the rod blank is set in the cradle of the winding jig. Winding silk is brought through the tension adjustment washers and four or five turns of the loose end of the thread are taken around the rod from back to front as depicted in sketch A.

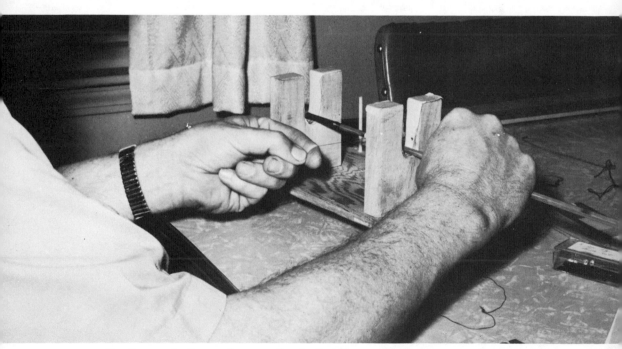

Starting a winding on homemade winding jig.

WINDING GUIDES AND KEEPER

(A)

(WINDINGS SHOWN WITH GAPS FOR CLARITY)

START OF WINDING

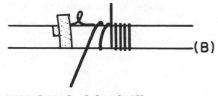

(B)

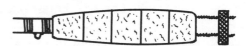

WINDING CAUGHT

KEEPER CLOSE TO CORK GRIP

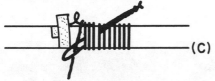

(C)

LOOP IN PLACE FOR FINISH

(D)

PULLING FREE END THROUGH

KEEPER LEFT SIDE OF ROD FOR RIGHT HANDED FISHERMAN

(E)

LOOSE END CUT OFF

IF SLIGHT BOW IN ROD SECTION, GUIDES PLACED ON OUTSIDE OF BOW

87

Maintaining some tension on the loose end of the winding silk with the left hand, rotate the rod counterclockwise with the right hand so that the supply end of the winding silk crosses that already on the rod. Several turns of the winding silk are applied and the loose end of the silk is cut off with a razor blade.

The winding is then continued to a point where four or five turns will complete it. Windings should be applied so that each successive turn just touches the preceding one with no overlaps or gaps in the winding. At first, the tendency here will be to go slowly and this is a mistake. For some reason, the winding silk will lay in much smoother and more easily if the rod is rotated at a fairly fast pace.

When the winding is completed except for the last four or five turns, a

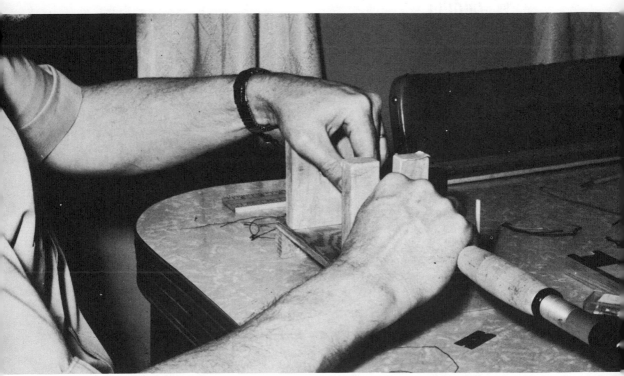

Applying winding.

previously knotted loop of winding silk is inserted under the supply end of the silk and the last four or five turns are completed leaving the loop extending beyond the winding, as shown in sketch C.

Holding the winding with the thumb of the left hand, cut off the supply end with a razor blade and thread it through the extended loop, as also shown in sketch C. Still holding the completed winding with the left

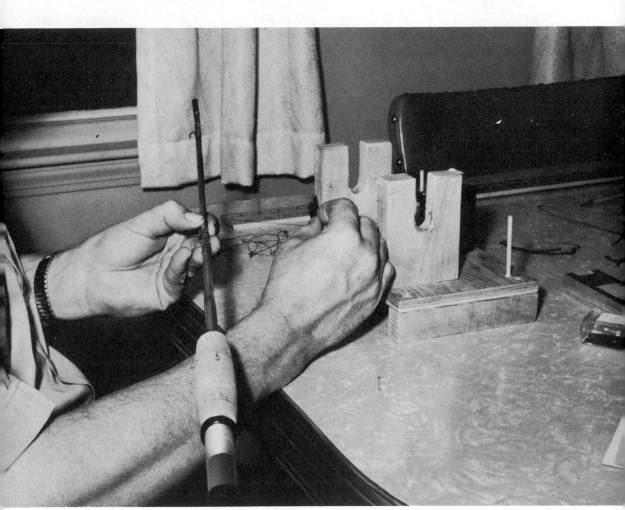

Winding held with thumb while "supply" end of silk is cut with razor blade.

thumb, pull the loop through under the last four or five turns of the winding, which will bring the free end back through the completed winding as shown in sketch **D**. The loose end is then cut off with a razor blade close to the completed winding and the process is complete.

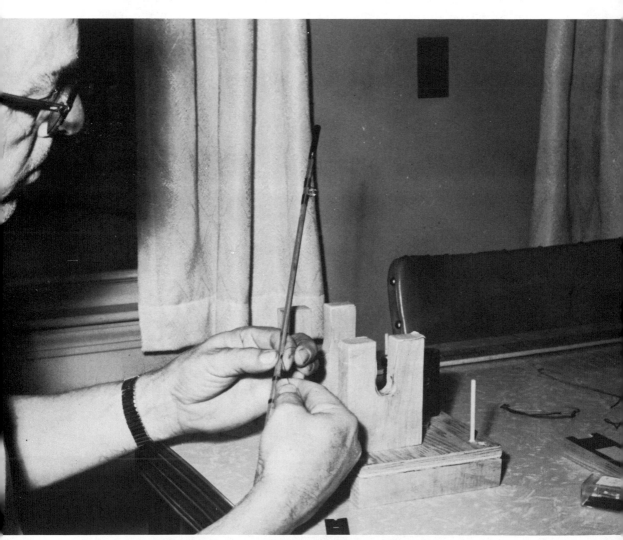

Free end of silk thread is tucked through loop wound under last 4 or 5 turns and is pulled through to complete winding.

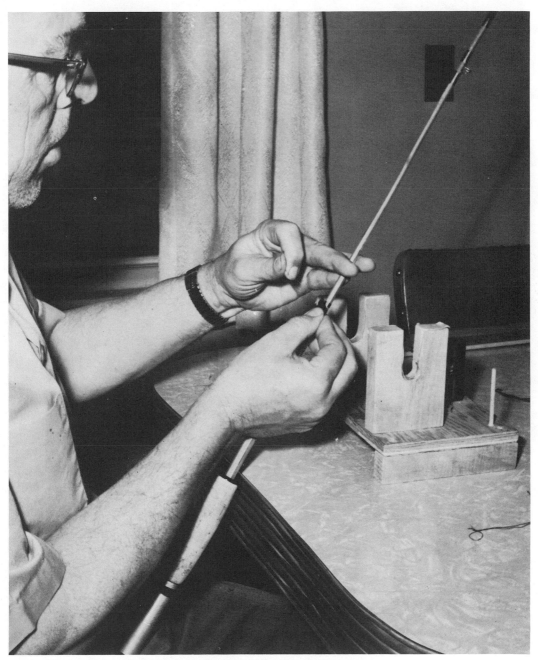

Free end is cut off close to winding with razor blade.

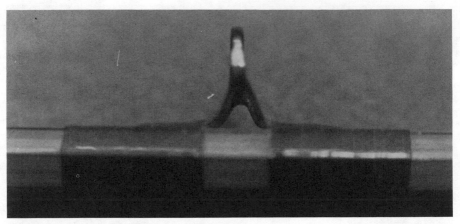

Closeup of stripping guide and winding.

The tape on the other leg of the snake guide is then removed and a winding is completed on that side of the guide.

The keeper should be located as close as possible to the tip end of the cork grip, as shown in the accompanying sketch. For the convenience of the fisherman the keeper should be located on the left side of the completed rod for right-handed fishermen and on the right side for those who are left-handed.

Similar windings are also applied at the rod end of each ferrule and are extended from the cane up onto the turned-down portion of the ferrule provided for this purpose.

At this point, if you cannot wait, it is safe to assemble the rod and lightly try its action. It should be remembered, however, that several steps still remain before the rod is ready for full action and fishing.

CHAPTER XV

COLOR
PRESERVATIVE
AND
FINAL VARNISH

A s soon as all windings have been completed, including those at ferrules, snake guides, keeper, winding check, etc., they should be given two coats of color preservative and six to eight coats of rod varnish.

Varnishing windings.

Color preservative, which is basically a lacquer, is applied to the windings with a small camel hair brush. Immediately following the first application, the winding is rolled between finger and thumb to help the color preservative penetrate the winding silk. A second coat of color preservative can be applied fifteen to twenty minutes after the first. The second coat of color preservative should be allowed to dry for several hours before varnish is applied over it.

Varnish is flowed onto the windings by rotating the rod blank. The process is the same as that used for varnishing the blank itself, except

Finished rods.

that the various coats are not rubbed with steel wool. If sufficient coats of varnish are applied, the windings will eventually develop the appearance of a smooth, highly glossed plastic tape.

CHAPTER XVI

FREQUENTLY ASKED QUESTIONS

A s I have become acquainted with more and more people who are undertaking rod construction for the first time, a pattern of recurring questions has developed. Most of these I think I have answered in the written material, photographs and sketches. On the assumption that further emphasis may still be desired, these questions, together with responses, are presented as follows:

Is there more than one source of supply for Tonkin cane?

Undoubtedly there are numerous sources of Tonkin cane. I have used only the source mentioned earlier and have been completely satisfied with both service and the quality of cane. The Thomas Register, which can be found in most libraries, should provide a ready reference for other sources of supply.

Are you as pleased with the fishing performance of your rods as you were with the rod-making process itself?

Yes. Possibly even more pleased. I do not class myself as a very proficient fly caster. There is no question in my mind, however, that my proficiency has improved with my own rods when compared to experience with commercial rods of both bamboo and fiberglass. Perhaps it is the double pleasure of fishing with a rod I built myself, or perhaps it is the compound rod tapers, or perhaps I am just more attentive to what I should be doing.

Do you still think resorcinol is a good choice of glue?

I have been more than pleased with the results of resorcinol. None of the joints have failed and it apparently has not deadened the action of the cane rod. The only possible objection, as stated previously, is the dark color of the glue joints.

Have you experienced any weak spots in rod construction?

On one rod I was not able to find a ferrule the exact size that I desired. Most ferrules come in $\frac{1}{64}''$ increments and I chose a ferrule $\frac{1}{64}''$ smaller, rather than going to a larger one. After fishing with the rod steadily for more than a week, I tried to reach a fish which had sounded beyond the normal range of my cast with this particular rod. When I applied additional power to the back cast, the rod snapped off at the point where the undersized ferrule had been installed. This, of course, broke off approximately an inch of the rod blank in question and it was repaired by turning a new ferrule seat and installing a larger-sized ferrule. Subsequent fishing with this same rod has not revealed any additional weaknesses.

How did you select line weight for the rods you have built?

I'm not at all averse to seeking advice from experts, and this is what I did when working on my initial rods. As each new model was completed, I took it to a friend who is the head of the fishing department in one of our major sporting goods stores; simply by flexing the fully assembled rod he was able to advise me on the proper line to use. His advice also included the emphatic statement that you should never buy an inexpensive fly line, as the casting quality is just not there. This may also be another reason why my fly casting has improved with my own rods, since prior to their construction I was prone to purchase a relatively inexpensive line.

Can you do rod construction with cane with an eight-inch node spacing?

Eight-inch node spacing in raw cane will leave only approximately 1¼″ between nodes in the completed rod and I feel that this is too close to provide the strength required. Cane with the nodes spaced at 10 to 13 inches will provide finished spacing of 1⅔″ to a little over 2 inches and this spacing is much more desirable.

If cane from two pieces of bamboo are used in the same rod blank, some problem with node spacing in the finished blank may be experienced, unless the raw cane has essentially the same spacing between dams in both pieces.

One or two articles I have read on rod construction advocate a node spacing of up to four inches in the finished rod blank. This has been completely impossible to attain with any Tonkin cane that I have used, but no problems have resulted with spacing in the neighborhood of two inches.

Is there any equipment I can use to straighten rod blanks after gluing?

At one time I toyed with the idea of building a tension device using either weights or threaded adjustments, but abandoned the project after talking with several other rod builders who had already tried the process.

I have not experienced any problem in straightening glued-up blanks by working them gently back and forth on the reverse side of the planing block, gently pressing them into one of the larger grooves on the block or removing major bends by eye.

Some care and patience must be used during the straightening process, if the rod is to be relatively straight. It is well to remember that a slight bow apparently has no effect on fishing quality so long as guides are installed on the "outside" of the bow.

Is there any way I can use or modify an electric drill to turn ferrule seats?

This may be a possible solution, but I have not been able to work out what I thought was a practical approach. One possibility might be to construct a blank somewhat longer than the finished rod section so that the very end of the blank could be chucked in the drill while a seat was turned down. Following this it would be necessary to cut off the additional length which was chucked in the drill. This process, of course, would preclude trying the ferrule on the seat and it would be necessary to take frequent and accurate measurements with the micrometer to insure a good fit. It would also be necessary to provide some type of support for the free end of the blank to prevent it from whipping.

When you split a piece of cane, how many rods can you build from it?

This is a question which has no answer. No two pieces of bamboo will be identical in size, shape or internal fiber arrangement. Some pieces will split better than others and more individual segments can be obtained from a piece which has good splitting characteristics than one with poor. The size of the rod being constructed will also dictate the number of segments which can be obtained from a piece of raw cane. Another unpredictable factor is the spoiling of a segment of cane while working it to final dimension.

Perhaps this problem was best summed up in correspondence with a professional rod builder in England when he said, "Working with a natural material means also that the 'rules' can never be the same. One has to adapt and change methods and find alternative ways to achieve similar results. Therein lies the fascination in rod building from bamboo."

Have you used any finish besides rod varnish?

I have finished one rod with a good quality of Spar varnish with excellent results. Comparison with rod varnish leads me to believe that it is nothing more than a good quality Spar varnish.

Have you modified your process of rod building as you gained more experience?

Each new rod has probably developed another trick or shortcut which has had no effect on the quality of the final product. Other than the feel of working with cane, which cannot be conveyed in writing, these shortcuts have been included in the narrative.

As an example of how quickly a little experience will affect speed of construction, my first rod took 53 hours to build; the second, which appears to be identical, took only 38.

Probably the biggest time saver that I have developed is heat treating completed rod segments rather than raw bamboo.

How do you know good rod cane?

Probably the honest answer to this question is that I do not. I have relied entirely on the supplier, making sure that he understood that I wanted the cane for fly rods.

Why are four grooves necessary in a planing block?

There is no magic in the number four. In working the cane, it is simply convenient to have a series of grooves, each one shallower than the previous. Three, five, or even six might be more convenient for another individual.

Isn't the slope of the planing block groove related to the rod taper, dictating a different planing block for each rod?

There is no connection between the slope of the planing block grooves and the rod taper. The grooves are simply an aid in working the cane to the proper central angle while taper is controlled by frequent and accurate measurements with a micrometer.

Do you move from one groove to the next in succession as a strip is formed?

This is my normal sequence of operation; however, some tapers may permit a groove to be skipped. With larger rod blanks, such as the butt end of a rod, it may not be necessary to use the finer groove in the block.

How can you be sure that a taper is uniform from one end of a rod segment to the other?

It would be unusual to find a rod with a uniform taper from one end to the other. Some inexpensive commercial rods may be built this way, but the majority of high-quality rods are built with compound tapers. One 6″ segment of a rod blank may be close to level while the next will have a sharp incline. On some rods—the Thomas Browntone, for example —a short steep taper will be encountered where the rod leaves the cork grip.

Can I use a scraper with replaceable blades available in most hardware stores?

I have yet to see a commercially built scraper sharp enough to work bamboo without tearing it.

Can I buy a scraper similar to the one you have used?

These scrapers are not available commercially and, so far as I know, have never been. The good quality of steel required for such a scraper and the frequent resharpening would preclude their manufacture as a standard item.

It is entirely possible to build a rod without the use of a scraper and several rod builders whom I have talked with advocate that they should not be used. I have found them very convenient in the final shaping, so long as they are extremely sharp.

Is the turned scraper pushed away from you or pulled toward you during final shaping of the rod segments?

The scraper is pulled toward you much the same as a commercial scraper with a wooden handle. If the scraper is properly sharpened and turned, the resultant shaving will literally be thin enough to read a newspaper through.

Doesn't it take more than normal home craftsmanship to build a rod?

My answer to this question is a definite no. Obviously, it takes care and some degree of patience to work with the fine dimensions that are required in rod construction, but it can in no way be classed as an impossible task.

Again quoting from my professional rod builder friend, "The truth is that anyone with some ability to handle tools and a desire to succeed can build an excellent rod at first attempt. Further trials and many errors later, better rods are built even by professionals."

CHAPTER XVII

THE
ULTIMATE TEST

A t this point, there is little advice that I can give, except assemble your reel and line and run to the nearest trout stream. There is nothing in the line of fishing that can match the thrill of that first strike on a handcrafted rod that you have built yourself.

I have tried to be explicit in all of the steps involved; however, if there are questions, I will be glad to answer them through correspondence. Obviously there are many variations of construction methods which can be used. The ones I have discussed may be no better than others, but I believe they will produce a more than acceptable rod.

In any event, I would enjoy hearing from you on how you made out with construction and your reaction to that first fish, the ultimate test of your rod.

Good fishing.

Index

Planing block
  in constructing rod, 42–44, 46, 47, 50–
      51, 52, 53–55, 58–59, 62, 68, 100
  construction, 6–7
  types, 2, 3, 6–7, 101–102
Planing strips, 50–51, 53, 55
Pliobond liquid adhesive, 29, 74, 76, 79
Preservatives, 29, 94, 95
Pressure wrapping bobbins, 22–23
Propane torch, 12, 13, 45, 47, 58

Rasp, 76
Rat tail file, 10, 12
Raw cane size, 38
Reel seat, 28, 68, 74, 76, 78, 79
Reel seat socket, 68
Resorcinol glue, 29, 65, 66, 74, 75, 76, 98
Rod bags, 30
Rod dimension chart, 54
Rod types
  8′ commercial rod, 32
  old Thomas rod, 32
  Thomas Browntone rod, 32–36, 102
Rough shaping cane, 38, 39, 42–44, 46
Rule, 11, 13

Sanding blocks, 11, 13
Sandpaper, 29, 53, 55, 68, 71, 76
Scraper, 10, 12, 16–18, 51, 68, 70, 102–103
Scraping strips, 51, 52, 55
Sharpener, 10, 12, 17
Shipping guide, 92
Six-strip rod, 2
60° center gauge, 10, 12, 51
Snake guides, 28, 86, 87, 92
Splitting strips, 38, 40, 42, 43, 44, 100–101
Steel planing block, 2, 3, 6, 7
Stick ferrule cement, 29, 79
Straightening cane, 38, 45, 46, 47
Straightening rod, 65–66, 99–100

Taper, 102
  Thomas Browntone rod, 32–36, 102
Tip of rod, 55, 86

Tiptop, 28, 78, 79
Tonkin cane, 28, 98, 99
Tools and equipment
  brushes, 11, 13, 29, 65, 82, 95
  burnishing tool, 11, 13, 17, 18
  carpenter's rule, 11, 13
  check list, 12–13
  "cutting" oil, 18
  finish winding jig, 23–25, 86
  gouge, 11, 13, 38, 40, 41
  hacksaw, 11, 13
  halving tool, 12
  heat treating oven, 20, 47, 58–60, 101
  knife, 11, 12, 38, 42, 47
  lathe, 68–72, 76–78
  leather gloves, 11–12, 13, 42, 50, 55, 79
  mallet, 11, 12, 38, 42
  mandril, 76–78
  micrometer, 11, 13, 32, 51, 53, 100
  mill bastard file, 10, 12, 17, 18, 40, 42,
      51, 55, 68, 72
  oilstone sharpener, 10, 12, 17
  pressure wrapping bobbins, 22–23
  propane torch, 12, 13, 45, 47, 58
  rasp, 76
  rat tail file, 10, 12
  sanding blocks, 11, 13
  sandpaper, 29, 53, 55, 68, 71, 76
  scrapers, 10, 12, 16–18, 51, 68, 70, 102–
      103
  60° center gauge, 10, 12, 51
  try square, 44
  wrapping vice, 20–22, 64
  see also Materials; Planing block
Triangular cardboard prism, 52
Try square, 44
Turning ferrule seats, 68–72

Varnish, 29, 82–83, 94, 95–96, 101

Winding check, 29, 77, 78–79
Winding jig, 23–25, 86
Windings, 28, 30, 86, 87–92, 95, 96
Wrapping thread, 59, 65, 66, 68
Wrapping vice, 20–22, 64